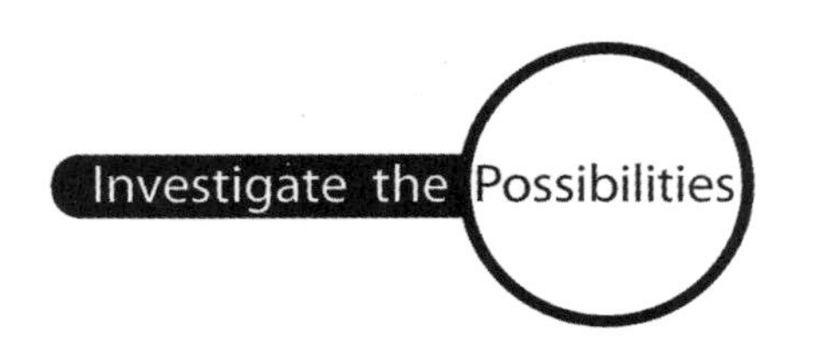

Elementary General Science

Tom DeRosa
Carolyn Reeves

Water & Weather
From the Flood to Forecasts
Student Journal

Tom DeRosa
Carolyn Reeves

First printing: October 2013

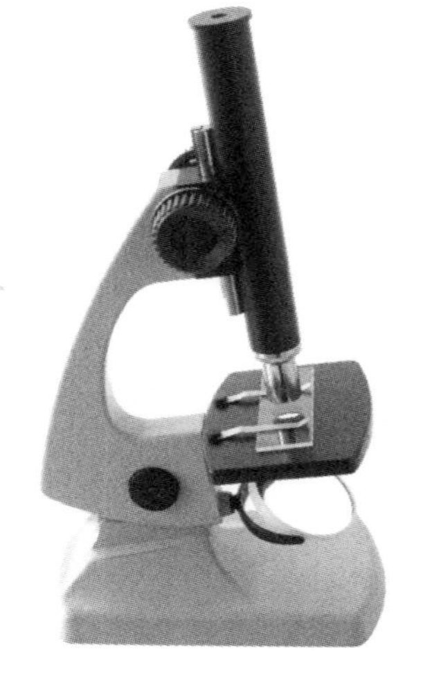

Master Books® is a division of the New Leaf Publishing Group, Inc.

ISBN: 978-0-89051-611-9

Cover by Diana Bogardus

Unless otherwise noted, Scripture quotations are from the New International Version of the Bible.

Please consider requesting that a copy of this volume be purchased by your local library system.

Printed in the United States of America

Please visit our website for other great titles:
www.masterbooks.net

For information regarding author interviews, please contact the publicity department at (870) 438-5288

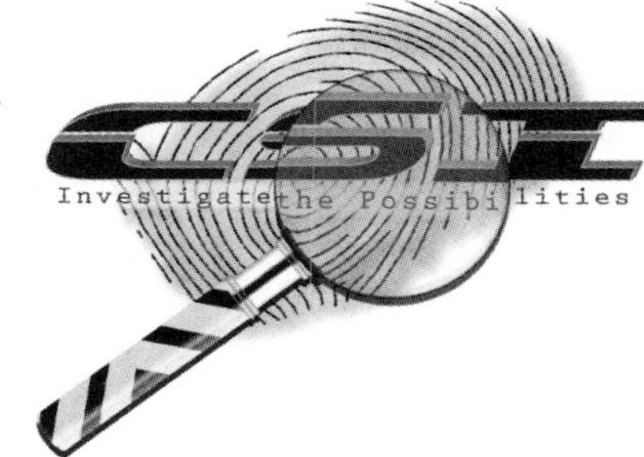

Table of Contents

Note to the Student

ecord your ideas, questions, observations, and answers in the student book. Begin with hink about This." After you read "Think about This," try to recall and note any experiences u have had related to the topic, or make notes of what you would like to learn.

ecord all observations and data obtained from each activity.

u should do at least one "Dig Deeper" project each week. Your teacher will tell you how any projects you are required to do, but feel free to do more if you find an area that is es-ecially interesting to you. The reason for the large number of projects is to give you choices. is allows you to dig deeper into those areas you are most interested in pursuing. Most of ese projects will need to be turned in separately from the Student Journal, but use the udent Journal to record the projects you choose to do along with a brief summary of each oject and the date each is completed.

Record the answers to "What Did You Learn."

The Stumper's Corner is your time to ask the questions. Write two short-answer questions related to each lesson that are hard enough to stump someone. Write your questions along with the correct answer or write two questions that you don't know and would like to know more about.

Some of these experiments should be done with the help of adult supervision. They have been specifically designed for educational purposes, with materials that are readily available.

ACTIVITY 1

Investigation #1

In the Beginning... God Created Dinosaurs!

Thinking About

Date:

The Activity: Procedure and Observations

Part A

Select a large dinosaur or an Ice Age animal and find the length of this animal from a reference book or the Internet. Make a beginning mark on the floor with a piece of masking tape. Measure the length of the animal with a measuring tape and mark its total length with another piece of masking tape.

1. Name the animal and tell how long it was.______________________

____

Cut several 3-foot strips of paper. Place the paper strips between the two pieces of masking tape.

1. How many strips of paper did it take to equal the length of this animal? ______________________

2. Describe the length in another way by estimating the number of cars that it would take to equal the length of this animal. ____________

____

Drawing Board:

Repeat this activity with a small dinosaur or Ice Age animal.

3. Name the animal and tell how long it was.______________________

____

4. Tell how many strips of paper (or fractions of the paper) it took to equal the length of this animal. ____________

____

5. Describe the length by estimating the number of cars (or fractions a car) that it would take to equal the length of this animal. ________

____

Part B

Place several smooth rocks inside a baggie. Add a small amount of vinegar to the baggie to simulate stomach acid, along with a piece of lettuce. Now rub the baggie to simulate a plant-eating dinosaur walki around with the rocks rubbing against the vegetation to help digest it

6. Describe what happens to the lettuce leaf in the bag. ____________

____

7. What do you think rocks do in a dinosaur's stomach? ____________

____

8. Why do you think vinegar was added to the baggie? ____________

____

Part C. Optional

Use reference sources and find the length of the strides of your dinos or Ice Age animal (from Part A). Compare the strides of the large and t small dinosaur/Ice Age animal. ____________

____

____

Dig Deeper

Stumper's Corner

What Did You Learn ?

1. Are dinosaurs classified as reptiles, amphibians, or mammals?

2. Did dinosaurs lay eggs?

3. Are woolly mammoths classified as reptiles, amphibians, or mammals?

4. Did woolly mammoths lay eggs?

5. Compare the sizes of the smallest and the largest known dinosaurs.

6. Give examples of dinosaurs that walked on two legs.

7. Give examples of dinosaurs that walked on four legs.

8. What are gastroliths and what seemed to have been their purpose?

9. Give examples of some Ice Age animals.

10. Where have the remains of huge numbers of woolly mammoth been discovered?

ACTIVITY 2

Investigation #2

Making a Big Impression!

Thinking About 

The Activity: Procedure and Observations

Part A.

Make an impression of a dinosaur skeleton model in layer of clay. Make a rim around the impression. Mix a batch of plaster of Paris and pour

Drawing Board:

plaster over the clay. Allow the plaster to harden, then carefully remov the plaster from the clay and observe the dinosaur impression. Look a the clay as well.

1. Describe the impression in the clay. ______________________

2. Describe how the bottom of the plaster of Paris looks. ______________________

Part B. Make a leaf mold.

While your plaster is still wet, put some of it in a small meat tray. Rub t leaf with petroleum jelly. Quickly, but gently, place a leaf on top of the plaster and cover with plastic wrap. Use a block to gently press the lea into the plaster. (Do not push down to the bottom of the tray.) After it has thoroughly hardened, remove the leaf. Compare the leaf to the lea mold. ______________________

Part C. Optional

Make molds of footprints walking and running.

Dig Deeper

Stumper's Corner

What Did You Learn ?

1. If something is pressed into wet sediment and becomes a fossil, what do we call the impression that is left behind?
2. What do we call anything that fills in an impression made by a living plant or animal and hardens?
3. What are several things a paleontologist might conclude about an animal that left fossil footprints?
4. Are small footprints in Coconino Sandstone examples of molds or casts?
5. What kind of clues do paleontologists look for in determining the size of an animal from its footprints?
6. Why do some scientists portray the Laetoli footprints as belonging to an ape-like animal, even though they clearly appear to be human footprints?
7. Are the results of radiometric dating tests of volcanic rocks always accurate?
8. What kind of fossils consists only of footprints or other impressions left by once-living organisms?

ACTIVITY 3

Investigation #3

No Bones About It!

Thinking About

Date:

The Activity: Procedure and Observations

Part A. Effects of different compounds on organic materials

Make three mixtures by combining and stirring the following ingredients:

¼ cup sugar plus 1½ cups hot water (stir until dissolved)
¼ cup salt plus 1½ cups hot water (stir until dissolved)
1 tablespoon of plaster plus 1½ cups cool water (stir until smooth)

Obtain three foam trays and label each tray — sugar, salt, or plaster. Place a sponge in the center of each tray.

Pour the mixtures very slowly over the each of the sponges. Place all three trays aside to observe for the next week.

1. Record your observations of the sponge soaked in the sugar mixture for a week. ______________________________

Drawing Board:

2. Record your observations of the sponge soaked in the salt mixture for a week. ______________________________

3. Record your observations of the sponge soaked in the plaster mixture for a week. ______________________________

Part B. Effects of vinegar and sodium bicarbonate on bones

Place one bone in a container of vinegar. Be sure to cover the entire bone. Place another bone in a container of water and 2 tablespoons o soda. Be sure to cover the entire bone.

Place both containers aside to observe next week.

1. Record your observations of the bone that had been soaked in vinegar for a week. ______________________________

2. Record your observations of the bone that had been soaked in baking soda and water for a week. ______________________________

Observe how flexible the bones are. Try to tie them in a knot. After you have finished making your observations, let them sit undisturbed in th air for another day. ______________________________

3. What happens to the bones after they sit in the air for several hours ______________________________

Dig Deeper

Stumper's Corner

What Did You Learn ?

1. In order for bones or other remains of once-living organisms to turn to fossils, what things usually happen?

2. When a dead plant or animal turns into a fossil, what role does dissolved minerals usually have in the process?

3. Which of the following is most likely to turn into a fossil — a dead buffalo lying on the ground or an animal that was covered by a large amount of sediment?

4. Is the formation of fossils a common occurrence today?

5. In one of the investigations you did, a bone was soaked in vinegar for one week. What chemical was lost from the bone as a result of a chemical reaction?

6. Explain why a massive worldwide flood would have provided ideal conditions for fossils to form.

7. Most fossils are found in which of the following kind of rock — sedimentary, volcanic, or metamorphic rock?

ACTIVITY 4

Investigation #8
Digging In and Reconstructing Fossils

Thinking About

Date:

The Activity: Procedure and Observations

Part A. Excavate chocolate chips from a cookie

Place a chocolate chip cookie on a paper plate. Using a toothpick and a brush, carefully try to pick out (excavate) all the chocolate chips. Be careful not to break the chips. Try to do this without picking the cooki up off the plate once you start.

Drawing Board:

1. How many whole chips were you able to remove? ______

2. How many were broken? ______

3. How long did this process take? ______

Part B

Your "fossils" have been placed in a "rock" by your teacher. Place the rock on a paper towel. Use a screwdriver or a sturdy wooden rod to ch away the rock until you see a hard object. This may be a real fossil or a model of a fossil. First, scrape around your fossil and then use a brush a toothpick to remove as much as you can to expose the fossil. Try not to scratch or break the fossil.

4. What problems did you encounter trying to find and remove the fossil? ______

Dig Deeper

Stumper's Corner

1. ______________________________

2. ______________________________

What Did You Learn ?

1. Why did most fossils form in sedimentary rocks? plant's or animals were buried by layers of sediment laid down by water.

2. Are all fossils found in a layer of hard rocks? no, some are found in sand.

3. Are fossils common all over the earth or is it rare to find a fossil? no they are comin.

4. Are most fossils the remains of sea-dwelling organisms or of land-dwelling organisms? sea dwelling.

5. Once scientists have excavated or collected a fossil, what information is kept with it? location and the envirонment wher it was collected.

6. Why is it often difficult for scientists to accurately reconstruct the skeleton of an animal after fossilized bones have been discovered? they are spred out in to pessis.

7. Do you think that most organisms that die and are covered in mud will eventually turn into a fossil? Give some reasons for your answer. they aren't bered depe anof

Investigation #5
Can Rocks Tell Time?

Thinking About

Date:

The Activity: Procedure and Observations

Measure 100 mL (a little less than ½ cup) of water in the measuring container. Add a few drops of food coloring to the clear glass. Take the styrofoam cup and punch a very tiny pinpoint hole on the bottom in the middle part of the cup. Once this is done, take the toothpick and place it inside the cup, carefully inserting it in the small hole that you have made. Leave the toothpick gently snug in the hole.

Place the styrofoam cup over the clear drinking glass to catch the water. Now gently pour the water into the styrofoam cup, trying not to disturb the toothpick. You should start to see drops of water come out of the styrofoam cup into the clear glass. The water in the clear glass w change color.

Feel free to adjust the toothpick by pulling it out until you get a steady rate of drops (about 10–30 per minute). Begin to count the drops and record the number of drops after a minute. Do this two more times, recording on the table below the number of drops that fall in one minute. The drops per minute should be approximately the same all three times.

Drawing Board:

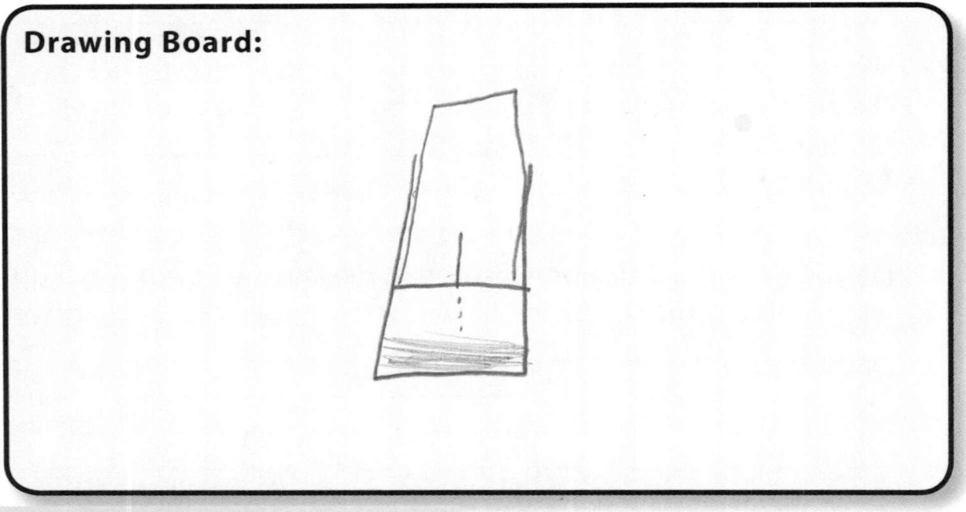

	Trial #1	Trial #2	Trial #3
Starting Time	0	0	0
Ending Time	1:00	1:00	1:00
Drops	34	40	41

Pour the water off from both containers and then start over again. For this part of the investigation you are going to measure the time it take for half the water in the styrofoam cup to drain off.

Pour 50 mL of water into the clear glass and mark this level with a rubber band. Pour out the water. Now repeat the experiment above, but this time stop when the water in the clear glass reaches the rubbe band marker (50 mL).

Use the toothpick to control the drops at the same rate as the drops pe minute in the previous investigation. Write the starting time and keep a close watch until the water level in the clear glass reaches the rubbe band marker. At this point write the ending time and subtract from the starting time on the table below.

Beginning Time 9:10

Ending Time 9:12

Time for 50 mL of water to drain 2:00

Dig Deeper

Stumper's Corner

What Did You Learn ?

1. The amount of time required for half of a radioactive substance to change into something else is known as what? half-life

2. In order to reach a more stable state, some radioactive atoms throw out little bits of nuclear particles, causing the original element to change into something else. This process is known as radioactive decay.

3. Scientists try to determine the age of certain things by determining the amount of radioactive substances.

4. When radioactive dating methods are used to date the age of certain rocks, do they always give the correct age of the rocks? No

5. Are carbon dating methods usually correct when dating the age of such things as wood or cloth? yes

6. Scientists assume that rocks containing both potassium and argon are the result of the radioactive decay of potassium into argon. What might cause the results of potassium/argon dating of a rock to be wrong? the argon did not come from the potacsium

7. Does absolute dating mean the same things as proven correct? No

ACTIVITY 6

Investigation #6
Leave No Stone Unturned

Thinking About 

Date:

The Activity: Procedure and Observations

This investigation will be a mental puzzle. As you have already learned, geologists, like detectives, have to work with clues and try to piece them together to figure out logical explanations for what might have happened in the past.

Part A – Study the diagram below and try to put these events in order from what probably occurred first to what probably occurred last. Make your best guess, but put a "?" by an event if you think there is not enough information to tell when it occurred.

a. sedimentary layers harden by processes of cementation and pressure
b. heavy rains erode a canyon through several sedimentary layers
c. sediments were laid down by water in flat layers
d. one section of sedimentary layers was pushed up (uplifted) by pressures below

Part B – Look at the diagram below and see if you can determine the correct order in which the rocks were laid down and cemented, as well as how other geological features fit in. You won't always be able to know for sure which is the older of two events, but you can use logic to provide a reasonable sequence of events. Sedimentary layers are identified by letters. Rearrange the following events starting with the oldest and going to the most recent: Sed L, Sed B, Sed D, Sed J, Sed Z, igneous intrusion, Sed A, Sed C, Sed F, Gully formed, uplift of rock (producing unconformity).

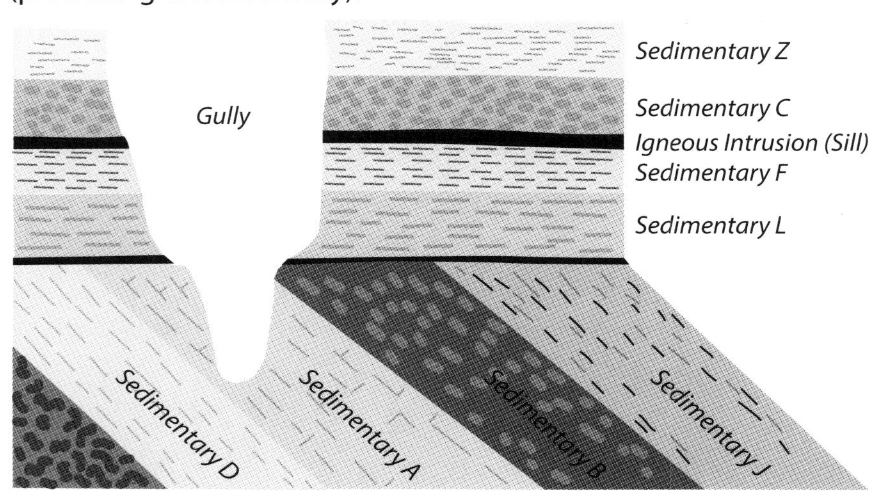

Now try to answer the questions below that refer to the diagram.

1. Which sedimentary rock layer is the oldest? ____________________
2. What is the youngest sedimentary rock layer? ____________________
3. Which formed first — the sedimentary rock layers or the intrusive rock layers? ____________________
4. Did the gully form before or after the sedimentary layers formed? ____
5. Did the gully form before or after the intrusive rocks formed? ______

Dig Deeper

Stumper's Corner

What Did You Learn?

1. What is the principle of superposition?
2. What is an unconformity?
3. What is meant by mass extinction?
4. Plants and animals that once lived in shallow parts of oceans tend to be found in lower layers of rock. How do evolutionists explain this and how do creationists explain this?
5. What does the geologic column supposedly show?
6. What are index fossils?
7. What does relative dating try to determine?

ACTIVITY 7

Investigation #7

Just How Salty is the Ocean?

Date: Feb, 1, 2017

The Activity: Procedure and Observations

Part A Add 4 ounces of distilled water to each of the three glasses. Leave the first glass of water as is, with nothing added to it. Add a pinch of salt to the second glass and stir thoroughly to dissolve the salt. Add ½ teaspoon of salt to the third glass and stir thoroughly to dissolve the salt.

Taste each glass of water in order — the distilled water, the water with the pinch of salt, then the water with the half teaspoon of salt.

1. Record what you noticed about the taste of the water in each glass.
1 good 2 bad 3 descosting

Drawing Board:

Thinking About

2. The glass with the half teaspoon of salt added to it is about the same as the amount of salt in seawater. Do you think you could you continually drink seawater? No

3. In your opinion, do you think ocean-dwelling animals could continually live in fresh water? No

4. In your opinion, do you think there are places where fresh water an salty ocean waters mix? yes

Part B Empty and clean the 3 glasses from Part A. Place 6 ounces of tap or distilled water in each of the three glasses. Place each glass on a sheet of paper. Label the paper according to the amount of salt in eac and add any other notes you need to keep up with. Do not add any sa or sugar to the first glass. Put 3 teaspoons salt in the second glass. Stir thoroughly. Put 3 teaspoons sugar in the third glass. Stir thoroughly.

Carefully place a wedge of potato into the first glass. Do not drop or toss it in. Observe.

1. Describe what happens to the potato wedge. it sank

Carefully place a wedge of potato into the second glass. Observe.

2. Describe what happens to the potato wedge. it flote bit

Carefully place a wedge of potato into the third glass. Observe.

3. Describe what happens to the potato wedge. it sank

dd a few spoons full of pure water to the salt water. Observe as the otato wedge begins to sink. Add a few more spoons full of pure water o the salt water.

. Record your observations of what happens. it sank

emove the potato wedges and place an ice cube in the pure water and n the salt water.

. Record your observations of what happens. they float and the salt floats higher

art C Pour 4 cm of corn syrup (or maple syrup) in a clear glass. Add 4 m of vegetable oil to the glass. Add a few drops of food coloring to 4 m of water and slowly pour the colored water into the glass. Make a iagram of the glass and label the layers of liquid in it.

arefully add a small piece of potato to the glass. Show in your diagram he layer to which the potato sinks.

dd a few other small objects, such as a rock, a cork, or an ice cube. bserve where each object sinks to or floats on, and show this in your iagram.

Dig Deeper

P—Potatoe i—ice T—Toothpick
C—crayon
e—eraser

oil
water
syrup
P
C
e
ROC

What Did You Learn ?

1. Which of the following has the greatest density — salt water or distilled water? Salt water
2. How is the density of an object determined?
3. Explain why the potato wedge in your investigation sank in pure water but floated in salty water.
4. When one solution floats on another solution, what does that tell about the densities of the two solutions?
5. What is an estuary? How does an estuary differ from open ocean water?
6. What happens to sediments that are carried by a river when it reaches the ocean?
7. If you caught a fish from the ocean and brought it home with you, what kind of environment would you have to give it in order to keep it alive?
8. Which is denser — ice or water? How can you tell?
9. Can a fish that lives in the ocean be kept alive in a freshwater aquarium?
10. Why do things float so easily in the Dead Sea?

ACTIVITY

8

Investigation #8

Ocean Zones – From Light to Dark Places

Thinking About 

Date:

The Activity: Procedure and Observations 

Take a clean empty milk carton that has three holes in the side and cover the holes with a piece of duct tape. Fill the carton with water. Put the carton in a large plastic pan or in a sink to catch the water.

Drawing Board:

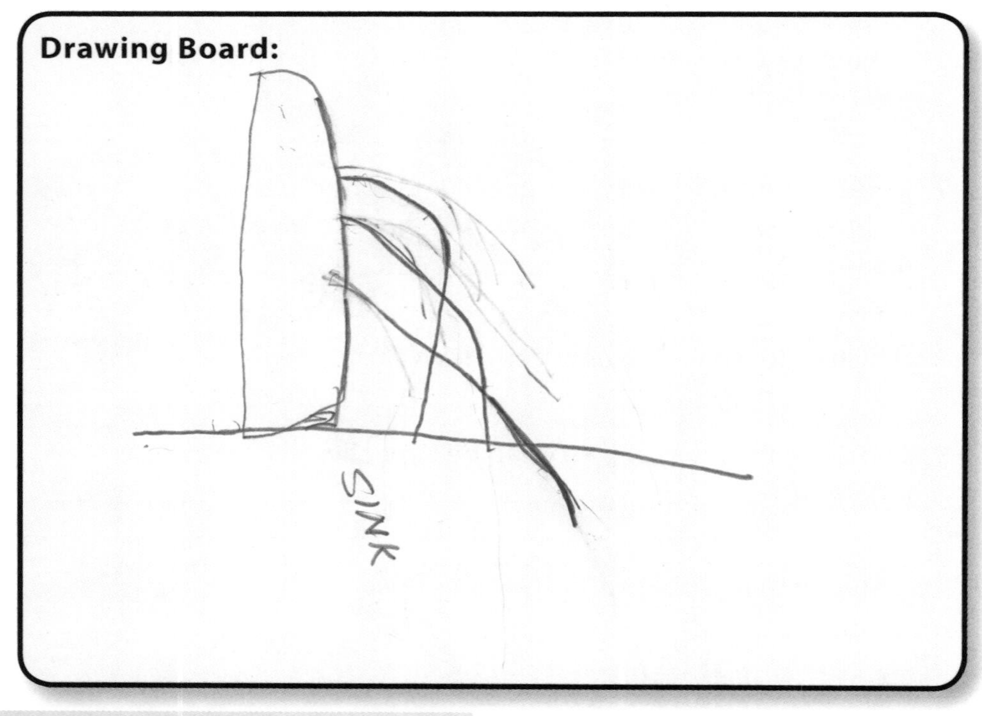

Before you remove the tape, predict what will happen when the tape is removed.

1. What do you predict will happen as water flows out of the holes?
   top is the least pressure middle not as strong botom strang top shoots less, middle more, botom most

Remove the tape and observe the water as it flows from the holes.

2. Describe how the water flows from the holes. Make a diagram to illustrate this. what I said in 1

3. Which hole had the weakest stream and which had the strongest?
   top botom

4. Did these results fit your prediction? yes

Dig Deeper

Stumper's Corner

1. ______________________________

2. ______________________________

What Did You Learn ?

1. Explain why there was a difference in the streams of water that escaped in the top and bottom holes of the milk carton. pressure

2. Explain why scuba divers don't usually go past depths of 130 feet (40 meters). too much pressure

3. What is the difference in the intertidal zone and the neritic zone of the ocean? intertidal zone covers low to high tides, neritic zone covers the continental shelf

4. In what ocean zone are most of the world's major fishing grounds found? neritic zone

5. What are some of the conditions that determine the zones in the ocean? pressure, sunlight, temperature, nutrients,

6. Are ocean plants and animals able to live anywhere in the ocean or do they only live in narrow zones where they are suited for specific conditions? no only zones where they are suited for

7. What are three kinds of habitats found within the intertidal zone of the ocean? Rocky Shores, coastal Wetlands estuaries,

8. Do different kinds of plants and animals live in the different kinds of habitats in the intertidal zone? yes

9. What important food source grows in the neritic zone in huge numbers and is a main source of food for many kinds of fish, especially the smaller fish, which may become food for the larger fish? plankton

ACTIVITY 9

Investigation #9

From Shelf to Shelf and In Between

Date:

The Activity: Procedure and Observations

Write a script for a TV interview with a scientist who has dived to the ocean floor in the submersible *Alvin*. Imagine you were the scientist aboard and your last mission was to explore some hot water vents and the living organisms around them. You were selected because you are an expert on the geological features of the oceans and have made two other trips in the *Alvin* to explore underwater ridges and mountains. Collect pictures made by *Alvin* to talk about. Describe the submersible and its capabilities. You will need to do a lot of research in order to do a good job on this. Begin your script with an introduction of the scientist (you) and the topic of the day. Here are some of the questions the scientist (you) will need to answer. You may change any of the questions below and you may write additional questions for the scientist to answer. Write your answers to the questions. Now get someone to be the interviewer and ask you the questions.

1. Can you show us on a map where the *Alvin* traveled during each of your missions? ______________________________

2. How long did it take for you to travel from the launching ship to the ocean floor? How long would your oxygen supply last? ______________________________

Thinking About

3. Describe the *Alvin* and the equipment on board. ______________________________

4. How much room did you have to move around inside the *Alvin*? ______________________________

5. What kind of information did you gather about the ocean during your first two trips? ______________________________

6. What kind of information did you gather about the ocean during the last trip? ______________________________

7. Could you show us some of the pictures you took on your missions? Tell us about the pictures. ______________________________

8. Many years ago, some scientists said it would be impossible for anything to live on the ocean floor. Why would they have thought this? ______________________________

9. Tell us something about yourself — what kind of training did you undergo before making this trip? Why did you want to study the ocean? ______________________________

(Alternatively, you can write a script about traveling in another submersible.)

? What Did You Learn

-4. Refer to the diagram of the ocean floor. Label the continental shelf, the continental slope, the deep ocean plains, and a seamount.

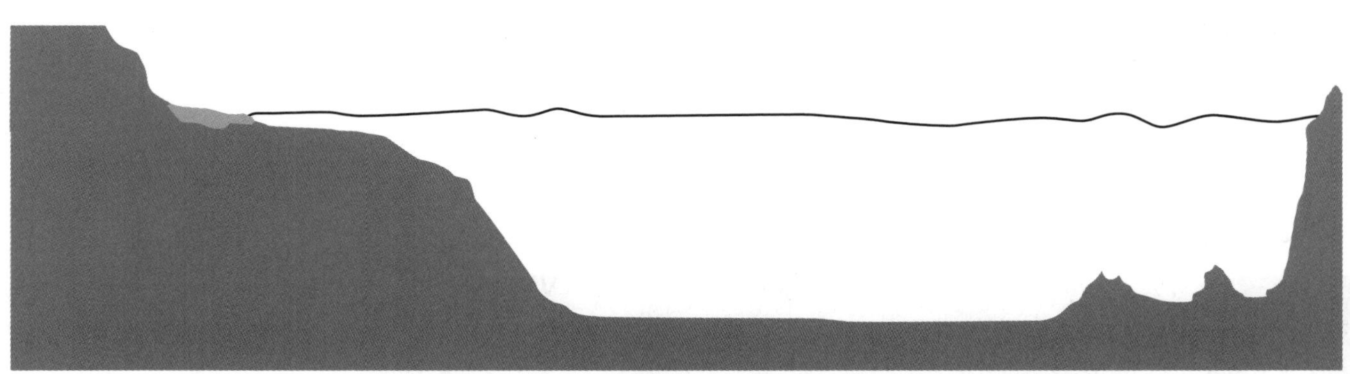

Dig Deeper

5. What is a submersible? Tell several things you have learned about submersibles. ______________________________

____

____

____

6. What are some of the geological features that have been discovered on the ocean floor between the continental slopes as a result of research done on submersibles? ______________________________

____

____

____

7. What are some of the conditions that exist on the bottom of the ocean floor, such as temperature, light, water pressure, etc.?

____

____

____

8. Since sunlight can only penetrate about 650 feet (200 meters) below the surface of the ocean, there are no green plants below this depth. What do animals that live there find to eat? ______________________________

____

____

____

ACTIVITY 10

Investigation #10

Currents in the Ocean

Thinking About

Date:

The Activity: Procedure and Observations

Be sure you have cubes of frozen colored water before you begin. Now fill a large pan with warm water from the tap. You will need to put one colored ice cube at each end of the pan, but before you do this, predict what will happen.

Drawing Board:

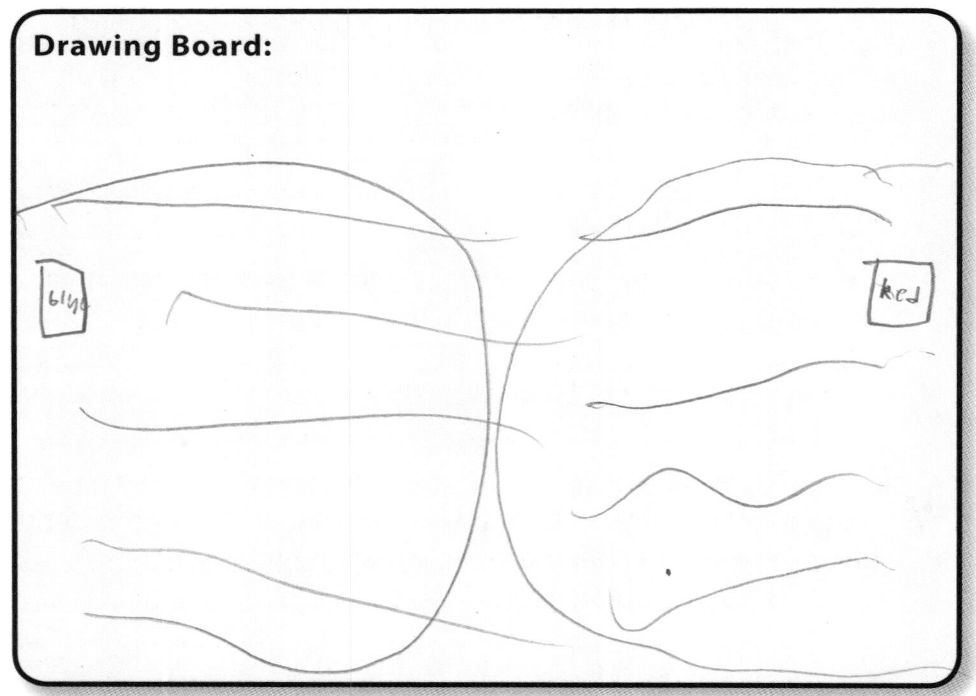

1. What do you predict will happen when you place a colored ice cube at each end of the pan of warm water. it will melt and then the food coloring

After you have written your prediction, place an ice cube at each end of the pan.

2. Write observations of what happens. it spreds and it meyd a line and the ice melt and wen thy ment it mayd a wall and they mixd

Lastly, place one drop of red food coloring in the middle of the pan.

Observe.

3. Write observations of all that is happening. ________

4. Draw what you saw.

5. As the cold water under the ice cube moved down, what water moved in to take its place? ________

Dig Deeper

Stumper's Corner

1. ______

2. ______

What Did You Learn ?

1. What is a convection current? Under what conditions do convection currents form in the air and in the ocean? Hot air rises cold air sink, when there are temperature differences

2. Explain how the Gulf Stream ocean current affects the climate of seaboard land areas in the United States, Canada, and Europe. warm air

3. What kind of pattern do surface ocean currents generally make in the Northern Hemisphere? clock wise

4. What are some of the things that cause or affect ocean currents? erth spinning densities of the ocean air currents

5. Explain why knowledge of ocean currents was important to early sailors as they crossed the oceans. So they wouldent get prosht

6. What important fact did Magellan's trip around the world tell others? connected

7. Both Ben Franklin and Matthew Maury were able to add to the world's knowledge about what subject? the oceans currents

ACTIVITY 11

Investigation #11
Where the Rivers Run Into the Sea

Thinking About

Date:

The Activity: Procedure and Observations

Make a copy of the map of the United States from the appendix showing major rivers and lakes and drainage basins. Use your reference sources to help you locate the following features.

Drawing Board:

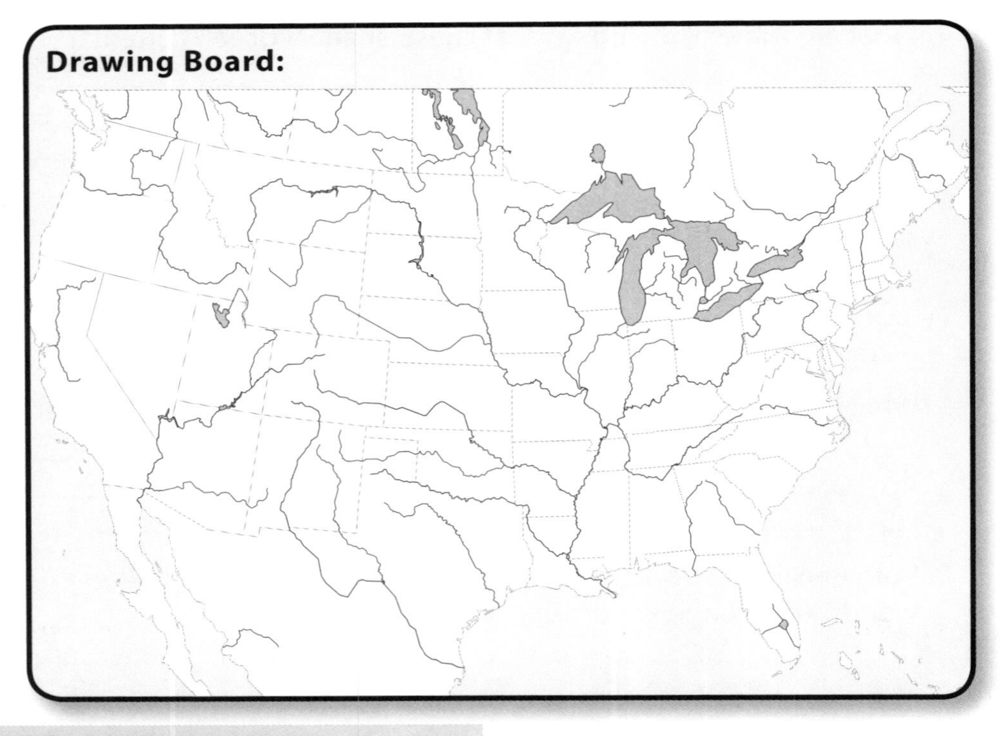

1. Color the Mississippi River in blue.
2. Color the major tributaries of the Mississippi River (rivers that empty into the Mississippi River) in purple. Label at least five of the major tributaries.
3. Find the area where the Mississippi River begins and write "headwaters" on the map.
4. Find the area where the river empties its waters into the Gulf of Mexico. Write "delta" on the map where this occurs. (Notice this area is divided into several branches.)
5. Trace the water route of Louis and Clark as they explored the Louisiana Purchase with a red marker. Use a black marker to show th part of their journey that was over land.
6. Draw an outline in green around the entire Mississippi River watershed, showing all the areas where water drains into the Mississippi River and eventually empties into the Gulf of Mexico.
7. Mark the Continental Divide in yellow.
8. Find the Great Basin and outline it in brown. Notice there isn't a plac where water from this area drains into an ocean.
9. Find the oceans into which each of the following rivers empties its waters. Notice that these rivers are not part of the Mississippi River watershed.

   a. The Colorado River ____________________

   b. Columbia River ____________________

   c. Rio Grande River ____________________

   d. Alabama/Mobile Rivers ____________________

Dig Deeper

Stumper's Corner

1. ______________________________

2. ______________________________

What Did You Learn ?

1. What is the Great Continental Divide? ______________________________

2. Could you start traveling by boat on one of the rivers on the east coast of the United States and connect all the way to the west coast by rivers? ______________________________

3. Does any of the rain and snow that falls in the Great Basin get transported to an ocean? What happens to it? ______________________________

4. Explain why the water in the Great Salt Lake in Utah is salty. ______________________________

5. Into what large river does water from the area between the Rocky Mountains and the Appalachian Mountains eventually drain? ______________________________

6. Does water from the Missouri River empty directly into an ocean or does it empty into another river? ______________________________

7. Have the Great Basin and the Sahara Desert always been a hot, dry environment? What evidence indicates that these places used to be wet environments? ______________________________

ACTIVITY 12

Investigation #12

Evaporation, Condensation, and the Water Cycle

Date:

The Activity: Procedure and Observations

Part A Dip a brush in a concentrated colored salt solution. Paint an arrow pointing up on the left side of a piece of paper and an arrow going down on the right side of the paper. Fan the paper to help the water evaporate more quickly. Examine the paper when it has dried.

Drawing Board:

Thinking About

1. What is the difference in your brush strokes when they are wet and when they have had time to dry? ______________________________
____

2. What was left on the paper after the water evaporated? __________
____

3. Where did the water go? ______________________________
____

After the paper has dried, use colored markers and draw a lake below the up arrow. Label the up arrow "evaporation." Draw a sun and a cloud in the sky above the up arrow. Write the word "condensation" in the cloud. Label the down arrow "precipitation." Draw trees and flowers below the down arrow.

Part B Add water to a shiny can until it is half full. Add ice cubes and stir. Observe what forms on the outside of the can.

4. What did you observe on the outside of the can that was not presen earlier? ______________________________

Set a bowl of water (half full) in direct sunlight or under a lamp. Cover the bowl with plastic wrap and place a small rock in the middle of the plastic wrap. Observe after several minutes.

5. What did you observe on the bottom of the plastic wrap? __________
____
____

6. Where did evaporation occur and where did condensation occur? (Find the definition of these terms if you need help.) __________
____
____

Dig Deeper

Stumper's Corner

1.

2.

What Did You Learn ?

1. Make a drawing of the water cycle and write a description to go with it.

2. What source of energy drives the water cycle? sun

3. When liquid water changes into water vapor, which process has occurred — evaporation or condensation? evaporation

4. When liquid water changes into water vapor, is energy absorbed or released by the water molecules? absorbed

5. Explain why rainwater does not contain salt even though much of the rainwater comes from salty oceans. when Water evaporates it leaves the salt behind

6. When a glass of ice water is placed in a warm room, drops of water will form on the outside of the glass. Does this water come from the air or from inside the glass? the air

ACTIVITY 13

Investigation #13

God's System of Purifying Water

Thinking About

Date:

The Activity: Procedure and Observations

Scoop up some dirt that has very small broken sticks mixed in with it. Add this to some clean water and stir well. Set aside a cup of the muddy water to use later.

Drawing Board:

1. Describe the muddy water. ______

Use the scissors to carefully remove the bottom of the 2-liter cola bottl **(Caution: Have an adult help you with this.)** Turn the bottle upside down, so that the pouring end becomes a funnel, and put the funnel end into a clean empty cup. Put coffee filters or a layer of cotton at the bottom of your bottle, covering the neck. Sprinkle water over this layer Add a layer of sand and sprinkle water over it until the sand is wet. The add a layer of small rocks. Now pour one cup of muddy water into the can and let the water trickle through the layers, directed through the neck into the empty cup.

2. Describe the water that was filtered through the sand and collected in the cup. ______

3. In your opinion, do you think the water is now safe to drink? ______

Dig Deeper

Stumper's Corner

1. ____

2. ____

What Did You Learn ?

1. Is most of the water on earth found as salt water or fresh water?

2. Where does rain water go after it falls on the ground and is soaked up by the ground?

3. Why is ground water usually clean and free of germs?

4. What are the advantages of filtering water that contains solid particles?

5. What kinds of things are not removed from water by filtering?

6. What are two ways of killing germs in water treatment systems?

7. Why are laws regulating garbage dumps important for keeping our water supplies clean?

ACTIVITY 14

Investigation #14
Weather or Not!

Thinking About

Date:

The Activity: Procedure and Observations

Part A

Observe the pictures of clouds below. If there are clouds in the sky, try to match them with the pictures.

Drawing Board:

Take three paper plates and write "cumulus" on one, "stratus" on one, and "cirrus" on one. Wear safety goggles to avoid getting shaving crear in your eyes. Squirt a mound of shaving cream on each plate. Shape the shaving cream on the first plate into forms that look like cumulus clouds. Do the same thing with the shaving cream on the other plates. Try to make the second one look like stratus clouds, and the third one look like cirrus clouds.

1. Make diagrams of the three basic cloud types and label each one. When you finish, show your teacher and make a note that you did this. ______________________________

A fourth common type of cloud is called a cumulonimbus cloud. Use a reference and find a picture or diagram of this kind of cloud.

2. Draw a cumulonimbus cloud on a sheet of white paper with a black crayon.

3. Use a reference to find an explanation for why "nimbus" is part of the name of this cloud. Find what kind of weather you would expec from this kind of cloud. ______________________________

Part B

You can make a fog/cloud in a bottle using the following method. Rinse a clean clear bottle (with a narrow neck) in hot water. Put about 2 inches (5 cm) of hot water in the bottle. Now put a large ice cube over the mouth of the bottle. Look through the bottle while the bottle is between you and the sun or between you and a light in the room.

4. Describe what you see. ______________________________

Dig Deeper

Cumulus Clouds

Cirrus Clouds

Stratus Clouds

What Did You Learn ?

1. Name the four basic types of clouds. cumulus stratus cirrus cumulonimbus

2. Which cloud type looks white and fluffy? Which one looks thin and wispy? cumulus cirrus

3. What kind of weather can be produced from cumulonimbus clouds? raine

4. What prefix can be added to a cloud name to mean the clouds are middle level? alto

5. Name two kinds of clouds that are not usually a sign of rain. cumulus cirrus

6. Suppose warm, humid air cools and condenses into water droplets that float near the ground. What weather condition forms? fog

7. Does warm humid air tend to stay near the ground or does it tend to rise into the air? rise

8. Explain how a cloud can form from a mass of warm humid air.

Investigation #15

Day and Night, Summer and Winter

Thinking About 

Date:

The Activity: Procedure and Observations

Part A

Hold your flashlight directly above a sheet of white paper (at a 90° angle). Shine the light from a distance of about 4 inches (10 cm). Outline the circle of light. Notice that it is in a circular shape.

Now hold your flashlight above the sheet of white paper at about a 30° angle, but still from the same distance above the paper. Draw around the light. Notice this time the light is not a circle, but is more oval.

Try to copy the two light patterns here.

Drawing Board:

Part B

Tape a string to the top of a styrofoam ball and put a piece of colored tape on the side of the ball. Dim the light in the room and shine a flashlight on the ball. Slowly spin the ball while the light continues to shine on it. This will represent day and night.

1. Write your observations of the ball as it spins. Tell when the colored tape would be in the daylight and when it would be in darkness.
2. Find the area on the ball where the light rays hit the ball most directly. Find the areas where the light rays are slanted when they h the ball.

Straighten a large paper clip and carefully push it through the center c a styrofoam ball. Turn the ball by holding the two ends of the paper cli The places on the ball where the paper clip extends represent the Nor Pole and the South Pole. Use a marker to make a line halfway between the Poles to represent the equator. Work with a partner and make careful observations of how the light shines on the ball.

Hold the ball by the two ends of the paper clip and tilt the ball about 20 to 25 degrees. Turn the ball so the "North Pole" is facing the light. Tr to determine the part of the ball where the light is most direct. Try to determine where the light hits the ball at a more slanted angle. Notice the difference in the amount of light that is shining on the "North Pole and the light that is shining on the "South Pole." This will be true even i the ball is spinning on its axis. (Refer to the diagram in the text to help you recognize these things.)

3. Make two sketches of the styrofoam ball held in this way with the light shining on it. Label one as "North Pole facing the sun." This represents the summer months. Label the other one as "North Pole facing away from the sun." This represents the winter months. In each sketch, identify the poles and the equator and show the lighte areas.

Dig Deeper

. In which position does the "North Pole" remain in the lighted area as the earth spins on its axis (turning the paper clips)? ______________

____

. In which position does the "North Pole" fail to receive light as the earth spins on its axis? ______________________________

____

. Describe the differences in the lighted areas on the ball when the earth is tilted and when it is not tilted. ______________________

____

What Did You Learn ?

1. What causes most countries to experience four seasons? the earth tilting

2. What is one complete spin of the earth called? How long does it take for the earth to make one complete spin? day, 24 hours

3. How long does it take for the earth to travel all the way around the sun? What is one complete path around the sun called? 365 days year

4. During which season of the year do the countries in the northern half of the world get the most direct sunlight? somer

5. Is summer in the United States the result of the earth being closer to the sun? no

6. What season do countries in the southern half of the world have when it is summer in the United States? winter

7. What two times are the effects of direct rays from the sun most noticeable? noon and morning

8. Compare the amount of heat energy the earth receives from direct rays from the sun with slanted rays of light from the sun. direct rays have more energy

ACTIVITY

16

Investigation #16

The Temperature's Rising

Date:

The Activity: Procedure and Observations

Part A Changes in Temperature of Dirt and Water

Put 1½ inches (4 cm) of water in one glass and 1½ inches (4 cm) of dirt in the other glass. Put thermometers at the bottom of each glass. Take the temperature readings of each glass. Place both glasses in full sun. (An alternative would be to place them under identical lamps.)

Record the beginning temperatures of the water and the dirt and then check and record the temperature again every 5 minutes for 15 minutes. (Try not to take the thermometers out of the dirt and the water to read the temperatures.) Record the temperature readings in the chart.

Move the glasses inside or to a shaded place out of the full sunlight. Check the temperatures of the water and the dirt every 5 minutes for another 15 minutes. Record these numbers in the chart.

	Time in Sun = 0	15 min	30 min	45 min
Dirt				
Water				
	Time Shade = 0	15 min	30 min	45 min
Dirt				
Water				

Thinking About

1. Use the numbers in your chart and make a line graph of your results for heating dirt and water. Use a red line for the water temperatures and a black line for the dirt temperatures.

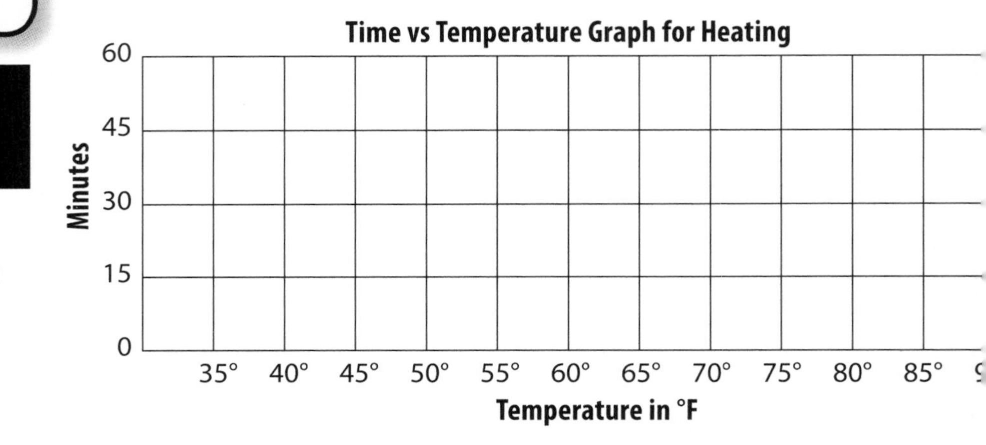

2. Use the numbers in your chart and make a line graph of your results for cooling dirt and water. Use a red line for the water temperatures and a black line for the dirt temperatures.

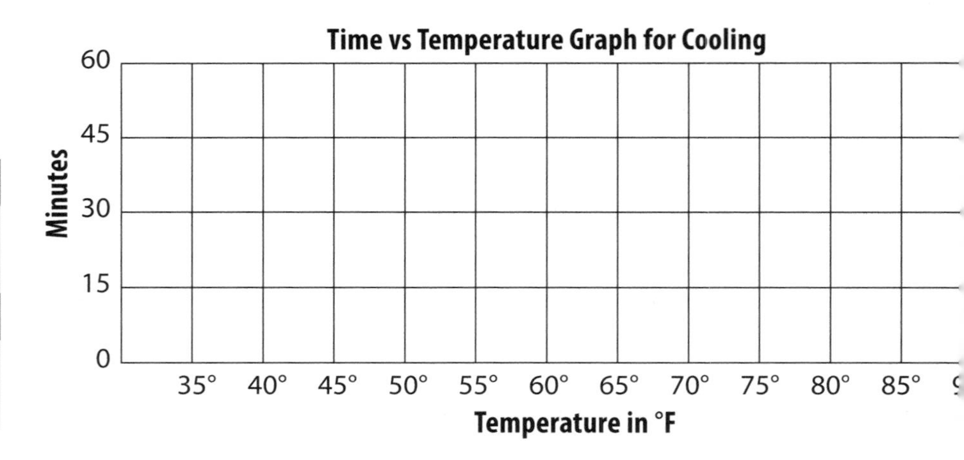

Dig Deeper

. Which substance heated up more rapidly? ____________

. Which substance cooled off more rapidly? ____________

art B

low up a balloon and rub ice cubes over it for about 30 seconds.

. Measure the circumference of the cooled balloon. An easy method is to put a piece of string around it, mark distance on the string, and then measure the distance with a ruler. Record this. ____________

low gently blow warm air over the balloon with a hair dryer for about 0 seconds.

. Measure the circumference of the warmed balloon as before. Record this. ____________

100000 0 000

What Did You Learn ?

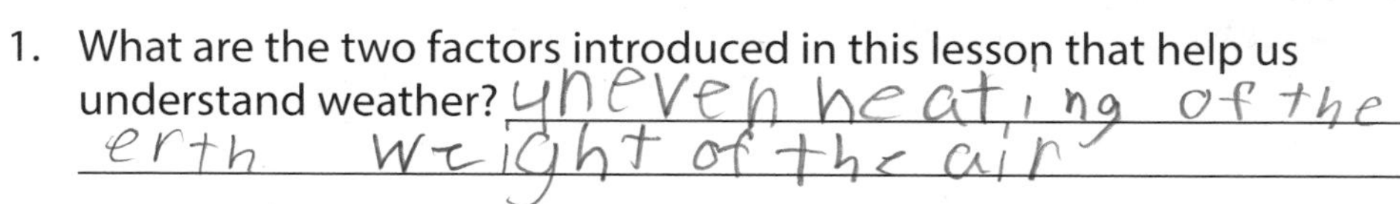

1. What are the two factors introduced in this lesson that help us understand weather? uneven heating of the erth weight of the air
2. Which one heats up faster — land or water? land
3. Which one cools off faster — land or water? land
4. Explain what happens to air molecules when they are heated. moove faster ferther apert
5. Explain how temperature differences in the air at different places cause wind. Hot air rise cold singkes
6. In what general direction do the prevailing winds over the United States blow? west to east
7. Suppose sunlight falls on a square meter of land at the equator and a square meter of land in an arctic region. Which square meter will absorb the most heat energy from the sun? Why is this? equator and the rase are more direcked
8. Would you expect the air in a weather "high" to weigh more or less than the air in a weather "low"? HI
9. Would a weather "high" be more likely to contain cold, dry air or warm, moist air? cold dry air
10. If there is a "high" over an area, what kind of weather would you expect to find there — fair or cloudy? sony
11. If you had one cold balloon and one warm balloon and both balloons were the same size, which balloon would be the heaviest? cold

Activity 17

Investigation #17
Weather Instruments

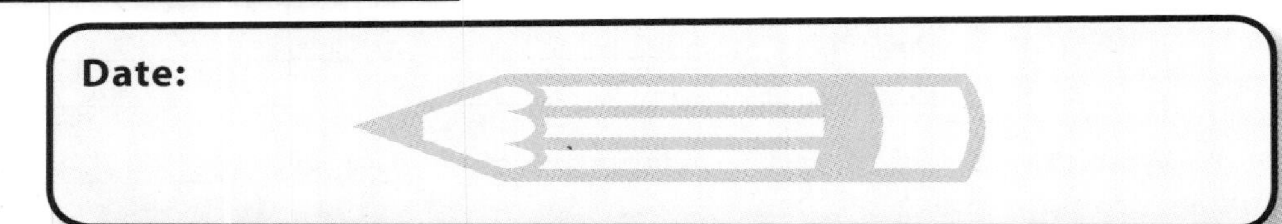

Date:

The Activity: Procedure and Observations

As part of this lesson, we will be building and using three common weather instruments and putting the information they provide into a weather chart. Use a regular thermometer to measure the temperature of the air each day.

Part A. Build three weather instruments and record weather information.

Build a wind vane

(Some supervision may be needed with the use of scissors and straight pins.)

Place a piece of thin paper over the "pattern for an arrow" in the appendix (page 44). Copy and cut out the pattern pieces. Place the patterns over a file folder or other thick, stiff paper. Trace around the patterns and cut out both pieces. Cut a 1-inch (3 cm) slit in each end of a drinking straw, and slide the front and back pieces of the arrow into the slits. Tape the arrow pieces securely to the straw, making sure everything is lined up.

Lay your arrow on your finger, and move it back and forth until it is balanced. This is the place around which your wind vane will turn smoothly. Mark the balance spot and push a straight pin through the straw from the upper side into a pencil eraser. Test your wind vane by holding the pencil and putting it in front of a hair dryer or fan. The trick is to make sure everything is lined up and balanced, so make adjustments if you need to.

Take your wind vane and a compass outside. Place the compass near the wind vane in order to determine the direction of the wind.

Record the direction from which the wind is blowing, as well as the da and time. You will be putting this information on the Weather Chart (page 44) each day.

A northerly wind blows from the north and a westerly wind blows fro the west. With the help of a compass, be as specific as you can. For example, if the compass shows that the wind is coming from between the north and the west, record its direction as from the northwest.

Build a simple barometer - Fill a clear, disposable plastic 1-liter bottl about 2/3 full of water. Add a few drops of food coloring. Quickly flip t bottle upside-down in a measuring cup. Some of the water may flow out of the jar into the cup. The right amount of water for the jar is a litt less than half full. If there is too much water in the jar, tilt it gently unti a bubble of air is released. Add additional water to the measuring cup until it is about half full. Use a permanent marker to show the top of th water level on the disposable bottle when it is in a vertical position.

Put a "1" by this mark. Each day that you observe the barometer, make another mark and put the next number beside it. In your notes, compare each number with the previous mark. Note if the barometer is rising or falling since the last observation. Record this information in your weather chart.

Build a rain gauge - All you need is a metal can, 6 inches (15 cm) or taller and a flat open space outside. Place a couple of bricks or rocks around it to keep it from tilting over. Don't place the can under a tree next to a building.

fter a rain, use a ruler to record the depth of the water in the can. ecord the depth in both inches and centimeters. U.S. weather reports sually give rainfall in inches. Check the rain gauge on the same ccasions that you check wind direction and barometric pressure. Pour ut any water after recording this information. If there has been no rain, ust record as "0." Snow has to be melted in order to measure.

art B

ollect information about weather conditions on at least five occasions. Jse the wind vane to find wind direction. Estimate if the wind is strong r mild (in your opinion). Determine if the barometer (air pressure) is ising or falling. Enter the information in the weather chart. Record the ime and the date when you collect the information. Also make a note n each of these occasions if the sky is cloudy or sunny. If you recognize he kind of clouds in the sky, indicate this as well. Make another note there is rain or snow. Each time you collect information from your omemade instruments, record the temperature of the air using a egular air thermometer.

fter you have collected weather information from all the sources and ecorded this information on the weather chart, predict the next day's veather. Tell if you were right or not.

Joseph Henry

Vhat were a few of Joseph Henry's accomplishments during his life? o you think he set worthwhile goals for his life? What are some of the oals you have for your life? ______

What Did You Learn ?

1. Is the uneven heating of the earth a major cause of weather conditions? yes
2. What two factors determine the air pressure in an air mass? water and tempetr
3. What does each of the following weather instruments measure?
   a. barometer air pressure
   b. thermometer temperature
   c. hygrometer humidity
   d. rain gauge amount of rain
   e. wind vane wind direction
4. If the surrounding air where you live is hot and humid, would you expect the barometric pressure reading to be high or low? Loy
5. If the winds blowing over Oklahoma City are coming directly from the north, would the temperature of the air be getting colder or warmer? coH
6. If the barometric reading at your home has been rising for the past day or two, would you expect the coming weather to be rainy or sunny? Sony

ACTIVITY 18

Investigation #18

Forecasting the Weather

Thinking About 

Date:

The Activity: Procedure and Observations 

Part A - Look at the weather map below and use the weather symbols in the appendix to help you identify the symbols. Label a cold front, a warm front, a high pressure area, and a low pressure area. The isobars are the lines around the highs and the lows. The pressure around them is given in millibars (mb) — 1013 mb is about average). Select the area near where you live and give as much information as you can about the weather conditions there. ______________________________

____

____

____

Drawing Board:

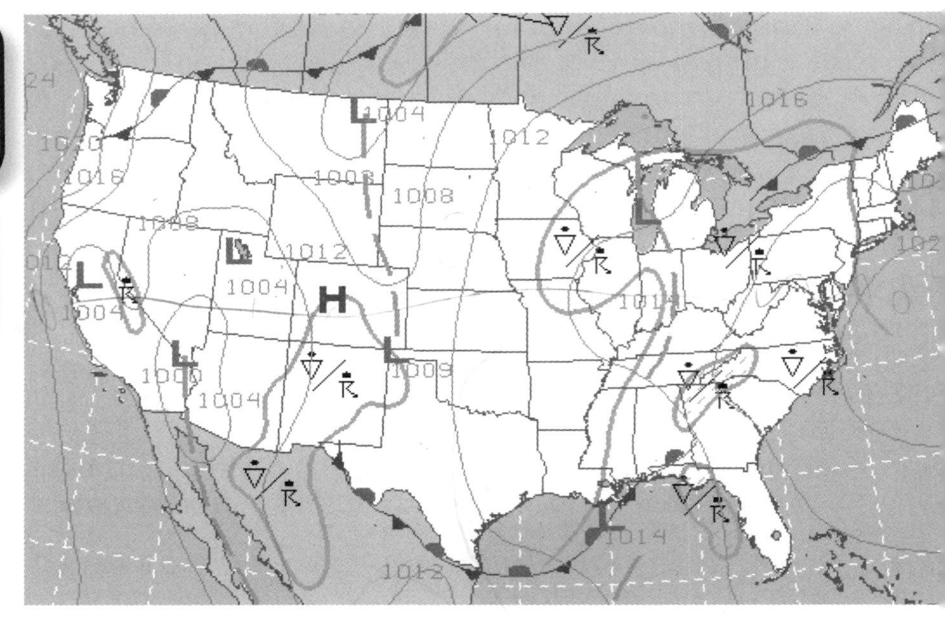

Some maps give additional information by using a kind of shorthand and other symbols. For example, the notations shown below indicate that the temperature is 45°F, there is moderate rain, the sky is overcast, the winds are from the southeast at 15–20 mph (13–17 knots), and the barometric pressure is 1004.5 mb. The dew point and other information is also often given on many weather maps.

Temp (F) Weather Dewpoint (F)	45 •• 29 045	Pressure (mb) Sky Cover Wind (kts)	**Data at Surface Station** Temp 45 °F, dewpoint 29 °F, overcast, wind **from** SE at 15 knots, weather light rain, pressure 1004.5 mb

In the space below, show the notations indicating these conditions: temperature of 34°F, there is light snow, the sky is overcast, the winds are from the northwest at 9–14 mph (8–12 knots), and the barometric pressure is 1003 millibars.

art B - Collect weather maps from newspapers or the Internet over five-day period. Get a copy of the first day's national weather map. ook for highs, lows, warm front, cold front, stationary front, clouds, recipitation, wind speed, wind direction, temperature, barometric ressure, and other information about the weather.

. Draw and identify the weather symbols you find on the map.

. Study the weather symbols and other weather information found on the map, and describe the weather conditions in three U.S. cities:

a. ______

b. ______

c. ______

. Sketch a city near the low pressure center. Mark the center of the low-pressure system with an "L." Draw arrows showing the direction of the wind around this low-pressure system. Include isobars of pressure if this information is given.

. Identify two U.S. cities from map 1 that probably have sunny weather and two that probably have cloudy or rainy weather. ______

ook at the map for day five and see if the weather conditions are still ne same for these cities. Were there any clues on the weather map that vould have helped you predict how the weather changed? ______

What Did You Learn ?

1. What are some indications that rain or snow may be coming soon? the foling presher and cold air mooving to wormair

2. What are some indications that sunny weather may be coming soon? High presher

3. What is meant by a front? leeding edge of a air mass

4. What are three kinds of fronts? Worm coll and stationary

5. Is the air following a severe cold front usually dry or moist? dry

6. What are some of the differences in weather highs and lows? ______

ACTIVITY

19

Investigation #19

From Gentle Breezes to Dangerous Winds

Thinking About

Date:

The Activity: Procedure and Observations

Remove the lids from both bottles and fill one with water to a few inches below the top. Touch the tip of the toothpick in some dish soap and just touch it to one of the small objects. Put it in the bottle with the water, along with any other small objects that can fit in the opening.

Drawing Board:

Put a little glitter in the bottle using the funnel, and add a few drops o food coloring. Put the empty bottle next to the one filled with water. Wrap duct tape tightly around the two bottles, so that there is a good seal around them. Turn the whole thing upside-down. Grab the top ar quickly spin it with a swirling motion. Watch the results.

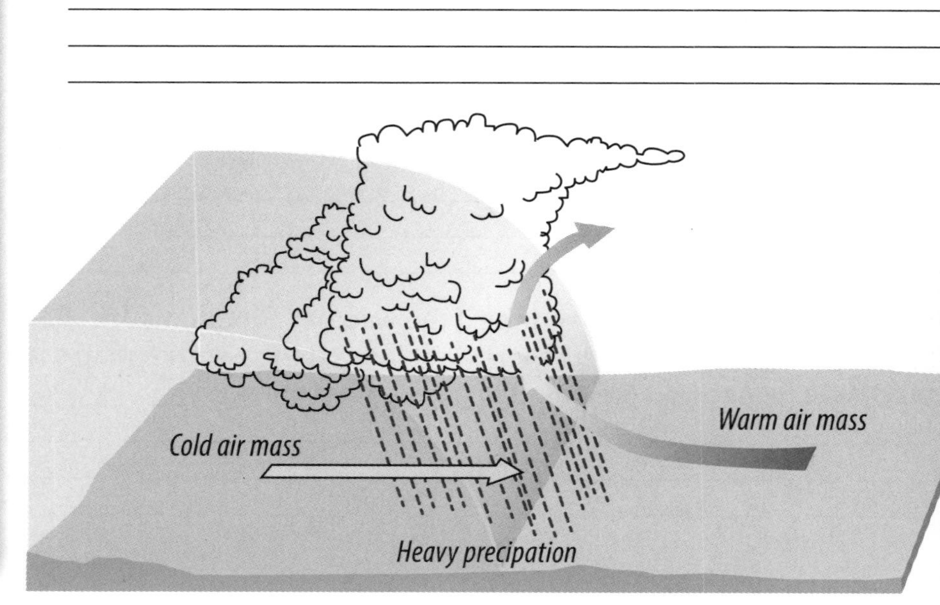

Dig Deeper

Stumper's Corner

1. ____________________

2. ____________________

What Did You Learn ?

1. What are the conditions that typically produce stormy weather involving high winds? Cold dry air mooves into worm moistare

2. What three things may occur as a mass of cold, dry air moves into a mass of warm, moist air? tornaco storm andwind

3. Does dangerous, stormy weather usually occur when a warm front moves into a mass of cold air? NO

4. Weather systems (including highs and lows) in the United States generally move from west to East.

5. How do the winds move around a low-pressure system? counterclockwise

6. Where do hurricanes that affect the United States originate? off the coast of Afreca

ACTIVITY 20

Investigation #20
Climate Changes

Thinking About

Date:

The Activity: Procedure and Observations

Freeze two sheets of ice in shallow pans that are the same size. The ice should be about an inch (2 cm) deep. Make a small opening in the top of a cardboard box and place the bulb of the lamp over the opening. Place one pan of ice inside the cardboard box. Turn the lamp on. Place the second pan in another covered cardboard box of the same size, but without a light bulb. Keep both boxes away from direct sun or a source of heat. Lay a thermometer inside each box about 2 inches (4 cm) away from each pan of ice. Record the temperatures after 1 minut and again after 15 minutes. Pour the melted water from each pan into measuring cup and record the amount of water in milliliters. Record al of your data in the data table below.

Drawing Board:

	Temperature after 1 minute	Temperature after 15 minutes	mL of melted water
Pan 1			
Pan 2			

Dig Deeper

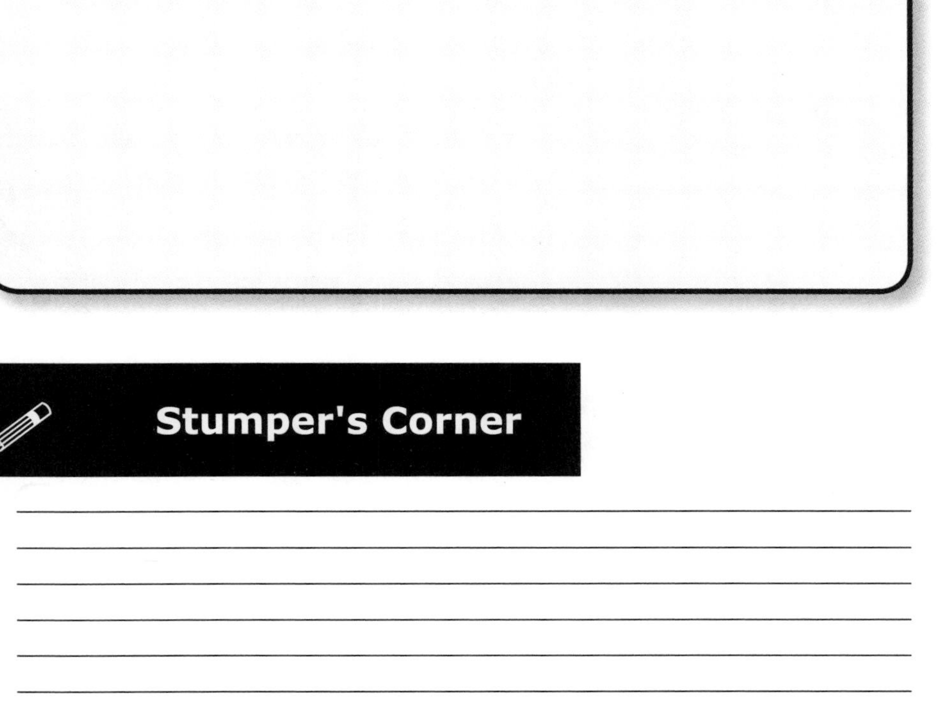

Stumper's Corner

What Did You Learn ?

1. What are two conditions that would be needed for an Ice Age to begin? more snow cooler somers
2. Explain why cold temperatures alone wouldn't cause an Ice Age.
3. What conditions might have occurred immediately after the Flood that could have caused the land to become cooler? volcanic ativity
4. What conditions might have occurred immediately after the Flood that could have caused the oceans to become warmer? What could have caused more rain and snow to fall during this time? valcano worm air over the oceans
5. What kinds of conditions were probably found in Siberia and other northern areas as the Ice Age began that were ideal for woolly mammoths? more food and water
6. Give an opinion about why you think many Ice Age fossils have been found in southern Florida. colder air wormer somers
7. Do geologists believe that sheets of ice covered a large portion of the United States during the Ice Age? yes

A Worldwide Flood

Joann was doing research on literature about early civilizations. The book she was reading explained that the earliest humans were extremely superstitious, because they were ignorant of scientific knowledge. They often passed on oral accounts of their heroes to their children. As people finally learned to write, these stories were written down and became the basis for myths. Stories about Greek and Roman gods and goddesses, fire-breathing dragons, and Noah and the Flood were all lumped together as "myths." As people learned more about science, they no longer believed superstitions and myths.

"Mom, tell me what you believe about Noah and the ark," Joann said as she looked up from her book. Mrs. Houston knew her daughter tried to gather facts and information as thoroughly as possible and didn't tend to make hasty conclusions. She was willing to examine information fairly. Mrs. Houston knew this question deserved a thoughtful answer.

"In all honesty, I can't give you definite answers about everything related to a worldwide Flood that occurred in Noah's day. However, I can give you some facts and some things that likely happened."

"The world that existed in the days of Noah was different from our modern world in many ways. For one thing, people tended to live to be hundreds of years old and remained healthy. Their long life spans allowed them to have large families with many children. They were able to accumulate advanced knowledge and skills. Most people had become so self-sufficient that they saw no need to honor God or live by His standards, even though God had created man to have fellowship with Him. They squandered their long lives and remarkable abilities on their own selfish desires without any regard for fulfilling God's purposes for their lives. There was much violence in the land."

"Do you really think there were giraffes, elephants, lions, bears, and other animals on the same boat for a year? Do you think there were any dinosaurs onboard?" Joann asked.

"We first need to be reminded that God is the Creator of everything, including the dinosaurs. The second point is that it would not have been necessary to bring adult dinosaurs or other large animals on the ark. Juvenile animals are small and could have easily fit in a small space. Another idea to consider is that many of the animals might have been in some kind of hibernation during their stay on the ark."

"During the time of Noah, there probably weren't extremes in temperatures like we see today on the earth, so special accommodations for hot and cold environments would not have been necessary," Mrs. Houston continued. "Many simil species we recognize today may have had the same ancestor. For example, I doubt that the ark contained two of ever kind of elephant and mammoth classified today as separat species. Both elephants and mammoths may have come from the same set of parents."

"The book I'm reading shows scenes of animals walking up a ramp two by two and entering the ark. It looks like a fairy tale or a nursery story. I wonder how much of that is realistic," Joann commented.

"The real story of the Flood is hardly an appropriate nursery story. The Flood was a terrifying judgment for the corruptio and violence that had filled the earth and even points to a future judgment of the earth by fire. Billions of fossils speak of the deaths of billions of animals."

Noah was about 500 years old when he began to build the ark. He was probably very intelligent and was likely an expert in animals and animal care, as well as in engineerin Also, God was very specific in the design, size and construction of the ark. The ark and the animals in the ark may

have been something like a modified zoo, but without the need for accommodating very cold environments. About a hundred years was invested in building the ark and providing places for animals to live. Don't assume that people in Noah's day were ignorant, lived without conveniences, or even that they wore long robes. In a culture where people lived hundreds of years, there would have been plenty of time to do research and gain knowledge. We really don't know what kind of scientific and technological advantages they had, since neither the Bible nor archeological records give us this information. What we do know is that some of the oldest cultures, such as Egypt, had remarkable engineering and mathematical skills."

"Let me ask you a question," Mrs. Houston paused for a moment. "Did Jesus consider the story of Noah and a worldwide Flood to be a myth or a real event? Listen to His words and see what you think He meant. 'As it was in the days of Noah, so it will be at the coming of the Son of Man. For in the days before the flood, people were eating and drinking, marrying and giving in marriage, up to the day Noah entered the ark; and they knew nothing about what would happen until the flood came and took them all away. That is how it will be at the coming of the Son of Man' (Matthew 24:37–39; NIV)."

"Jesus certainly acknowledged the Flood as a real event," Joann admitted. "However, I still don't see how a 40-day rain could cover the whole earth with water. Does that sound reasonable to you?"

"The water did not come only from rain. The Bible speaks of two sources of the water that covered the earth, both beginning on the same day. (1) All the springs of the great deep (the oceans) burst forth, and (2) the floodgates of the heavens were opened. This is recorded in Genesis 7:11–12."

"One possibility for what happened is that tectonic crustal movements caused the earth's crust to fracture, suddenly releasing a massive amount of very hot water that was stored in the crust. This would have sent steam into the air, which would have fallen as rain.

"The Genesis account continues with very specific information about how long the waters covered the earth. It tells that after a period of time, the springs of the deep and the floodgates of the heavens were closed."

"These ideas make sense to me, but I have two more questions," Joann said. Did the Flood waters really cover the tops of high mountains like Mt. Everest? And, where do you think all the water went?"

"There is much about the pre-Flood land we don't understand, but the oceans may have been somewhat shallow and the land may have been flatter with low mountains. The formation of high mountains and an uplifting of many other places on land probably occurred during or after the Flood, along with deep trenches in the oceans. The formation of deep ocean trenches would have provided deep basins for ocean waters."

"I'm glad we had this conversation," Joann concluded. "I still have questions about what happened, but I don't believe the biblical account of the Flood belongs in the myth category. And it is certainly not unreasonable."

What questions or thoughts do you have about the Flood?

A Worldwide Flood

Read this story and respond with some thoughts of your own. ________

____

____

____

____

____

A Worldwide Flood

Note

Students need to see the biblical Flood as a real event in human history. Secular science books do not even consider such an event. However, it is completely logical and provides a more satisfactory explanation for fossils and sedimentary layers than uniformitarian explanations. Noah and his family were real people, but they had amazing physical and mental features and lived to be very old. Encourage students to write their thoughts about the Flood.

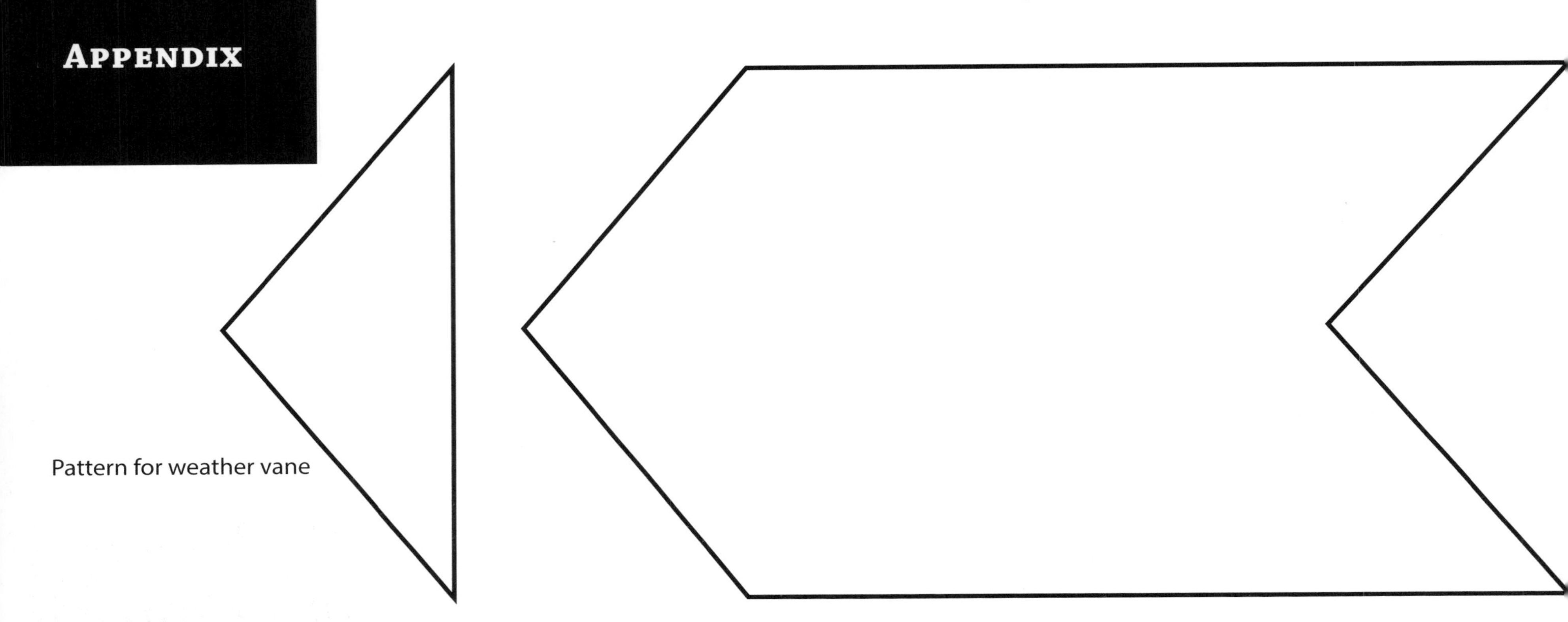

Pattern for weather vane

Weather Chart and Forecast

Date/Time	Temp	Wind Direction	Strong or mild Winds	Rising or Falling Barometer	Rain or Snow in Inches	*Cloudy or Sunny	Next Day Forecast	Were You Right?
1st								
2nd								
3rd								
4th								
5th								

*If you recognize the kind of clouds in the sky, include this. Otherwise just indicate whether the sky is sunny of cloudy.

Common Weather Symbols

Sky Cover	Wind		Fronts	Symbols
Clear	Calm		Cold Front	Rain
Scattered	8-12 Knots (9-14 mph)		Warm Front	Thunderstorm
5/8	13-17 Knots (15-20 mph)		Stationary Front	Snow
Broken	18-22 Knots (21-25 mph)			Fog
Overcast	48-52 Knots (55-60 mph)			

Additional Notes

United States map with rivers